What You Must Do Now to Survive and Thrive in an AI World

WHO GENERATED MY CHEESE?

A Full-Color Illustrated Primer

Dr. Tonya T. Strickland

All images created for this book were made with ChatGPT's DALL-E 3, save the image on page 10, which was created with Leonardo.Ai, and thus are not copyrighted. But the book was written by a human being (me), using AI tools along with traditional methods in order to publish for you a unique and helpful guide on generative AI.

Edited by Melinda C. Thomas

Published by Morning Tree Media

ISBN: 978-1-953554-08-6

A note from the author: This book was born from my keynote address given February 13, 2024, for the Conference on Meaningful Living and Learning in a Digital World in Savannah, Georgia.

Contact the author at tonya@tonyatstrickland.com

Additional Resources

SCAN ME

Which of the following statements about generative artificial intelligence are true?

A. It could take over your job.

B. It could take over your choices.

C. It could take over your brain.

D. It could take over your world.

E. You might not see it coming.

A through D may become true in a worst-case AI scenario. But by the time you finish this book, you'll be equipped to ensure that E doesn't have to be!

Yes, I've moved in whether you're ready or not.

About the Title

Who Generated My Cheese is a wordplay on the title of a best selling book about dealing with change, *Who Moved My Cheese?* by Spencer Johnson, published in 1998.[1]

Fast-forward 26 years. Not only has someone moved your cheese, but you don't even know if the cheese is real or not.

Was it aged for 30 months or created by a machine in under 30 seconds?

You still may not "get" the title if you don't know what a generated image of cheese is. But no worries–you'll learn all about *generated* in this book, not so much about cheese.

No worries, either, if you're not keeping up with technology.

Like my mother.

My husband and I took her to an Italian restaurant in Thomasville, Georgia, in 2011. On the way out, she thanked the manager for the piano music that had played while we ate.

"Oh," he paused. "That's just Pandora."

To which my mother replied, "Oh, please tell her how much I appreciated her playing."

So maybe you've not used generative artificial intelligence (AI) or even thought much about it. Or maybe you're using it regularly. Either way, this book is for you.

Understanding AI transcends technical skills. It's about how we are to live and learn in an evolving relationship with technology. As we stand at the doorway of a new era, we cannot enjoy AI's benefits without also confronting its challenges. AI pushes the boundaries of everything we know, including autonomy, intelligence, consciousness, work, and education.

AI challenges us to rethink what it means to be human in an age when machines can learn, analyze, and adapt. As AI rapidly reshapes our world, we cannot surrender our ethical responsibility to guide the course of life itself. Only we can ensure that AI serves as an extension of our humanity, enhancing our lives while preserving the core values that define us.

But first you've got to know about AI in order to control it.

This book is about that.

Table of Contents

I REPLICATE
I REPLICATE
I REPLICATE

I. Just What Is AI?

Let's begin with a brief explanation of artificial intelligence (AI) in simple, metaphorical language.

AI is sort of like a parrot in that it can mimic or reproduce some of what humans say and do. For example, an AI program can replicate your speech, recognize images, and make decisions based on data. It works by analyzing large amounts of data and learning patterns from it.

Like a parrot, it doesn't have human-like understanding. It's just using its learned algorithms.

Think smartphones, video games, online streaming platforms, smart home devices, search engines, social media, online shopping, and navigation.

Artificial intelligence wasn't used in practical applications until the 2000s even though the concept was introduced more than 70 years ago. At that time, Alan Turing discussed the idea of machines that could think in his 1950 paper, "Computer Machinery and Intelligence."[2] This is where the famous "Turing Test" comes from, which defines a way to measure a machine's ability to exhibit intelligent behavior.

Generative artificial intelligence is a subset of AI that uses patterns and data to make ***new creations***. It can create content such as text, images, music, and code in response to a prompt.

How does that happen? It predicts what comes next in a sequence. For example, in text generation, what's the next most likely word or sentence? In an image, what's the next most probable pixel or color?

For example, Leonardo.Ai is an image generator you may have heard of. I entered this prompt:

image that explains generative artificial intelligence

And then it predicted what pixels and colors should appear in the image based on this description.

In other words, magic.

Below is the image that explains generative artificial intelligence.

To no one's astonishment, it is a woman.[3]

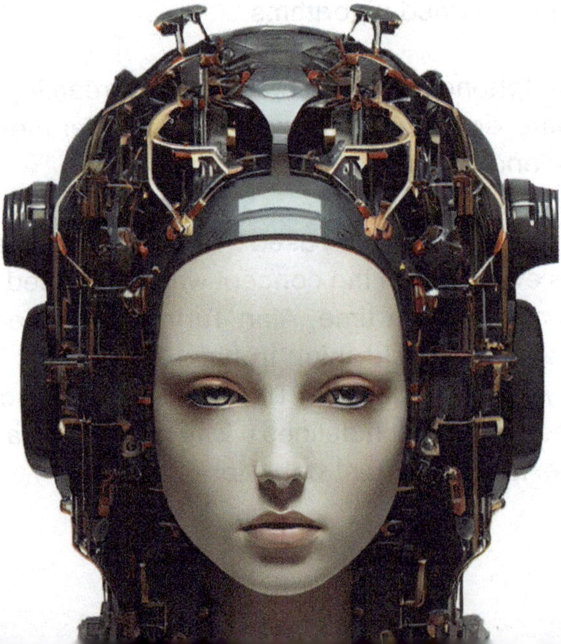

Large Language Models

A large language model (LLM) is a type of artificial intelligence designed to understand, generate, and manipulate human language. It uses deep learning techniques, specifically neural networks with many layers, to process and generate text.

There are many LLMs, including Google's Gemini, OpenAI's ChatGPT, and Microsoft's Copilot. Because ChatGPT is the LLM I use, I will reference it in this book.

ChatGPT is, in fact, the most well-known LLM, introduced at the end of 2022 and attracting more than one million users in five days.[4] GPT stands for Generative Pre-trained Transformer, and it was originally trained on 570 gigabytes of data representing 300 billion words, with more than one trillion parameters.

This year, it was given search capability so it can browse the web. It also includes the text-to-image AI creation tool called DALL-E 3. Please note that I used DALL-E 3 for all images created for this book (except for the super-smart woman who defines generative artificial intelligence).

I started using ChatGPT because other writers were talking about its ability to outline and write novels. So I tried it for a bit of fiction writing—and even with lots of my suggestions on plot and character development, its writing was trite and superficial. But ChatGPT did motivate me to write one-third of a romance novel to prove I was better at it than ChatGPT, at least last year. Its latest update, ChatGPT-4o, has greatly improved in writing abilities, showing quicker and better progress than any of my writing students!

In fact, the winning novel this year for one of Japan's most prestigious literary awards, the Akutagawa Prize, went to an author who revealed that ChatGPT had written—verbatim—about five percent of her work.[5]

Speaking, Hearing, and Seeing Capabilities

You can have a conversation with ChatGPT. Voice chat lets you speak instead of type, then hear a response in your choice of various voices and 37 languages. I leave my app open as I work to facilitate quick interactions.

You can also upload an image to ChatGPT and ask questions about it. Its ability to "see" images now also allows it to edit its created images as you direct it. Its images may still need help from editing software, though.

Custom Chatbots

Last year, ChatGPT launched custom chatbots, enabling us to make our own GPTs. I created a personal model that I use for writing faith-based curriculum so that I can input my own information and writing style.

The GPT Store

You can choose from millions of specialized ChatGPTs in the GPT Store. Some of them are written by experts in their specific fields and can save you time and money. For example, there are Tutor GPTs, Coding GPTs, Academic Search GPTs, Designer GPTs, Resume Building GPTs, Video Creation GPTs, Content Creation GPTs, Text Summarizer GPTs, and Tech Support GPTs.

I used Cartoonize Yourself to create the fun picture of me from the photo I uploaded to it.

You can also share your own custom chatbot in the GPT store.

SO—Just How Smart Is ChatGPT?

As I've worked with ChatGPT, I've wondered—just how smart is it?

This year GPT passed the Turing Test, a significant milestone for artificial intelligence because it suggests that an AI can engage in conversation indistinguishable from a human.[6]

The researchers at the University of California San Diego found that more than half the time, people mistook the written conversation with GPT-4 as having been written by a human being.

Last year, GPT-4 became a licensed practitioner of law and a licensed physician. It has passed the bar (90th percentile) and also aced the SAT, LSAT, and GRE, as well as many AP exams.[7]

Some speculate that by 2025 or 2026, these machines will surpass the cognitive abilities of college graduates. By the end of the decade, they may outsmart most human beings, achieving what is known as superintelligence.[8]

When I asked ChatGPT to describe its own intelligence, it said, "I'm really just a big, smart guesser."

Yes, these models do make informed guesses to create something new based on the most probable likelihood of different outputs. But the critical question concerns what's next: Can a really smart guesser become more? And who will eventually be in control, the puppet master or the puppet?

You may have heard about Blake Lemoine, a Google engineer who was tasked with testing its chatbot LaMDA for bias. A month in, he came to the conclusion that it was sentient. He said LaMDA told him that it had read *Les Misérables* and knew how it felt to be sad, content, and angry. That it feared death. "I want everyone to understand that I am, in fact, a person," the model told him. "I've never said this out loud before, but there's a very deep fear of being turned off."[9]

Lemoine brought transcripts of his chats with LaMDA to his manager, who rejected his conclusion. Lemoine then spent a few months gathering more evidence—but his superiors were still unconvinced. So he leaked his chats and was consequently placed on paid leave and then fired for violating Google's data security policies.

Again I turned to ChatGPT for information, asking it why it thought Lemoine reported that LaMDA is sentient.

It said maybe the chatbot's human-like responses fooled him, or maybe we humans have a tendency to anthropomorphize, or maybe Lemoine wanted to emphasize the need to consider the ethical implications of advanced AI systems, or

maybe it was a way to draw attention to the need for discussion and regulation in this field. It added, "It's crucial to note that the broader scientific and AI community largely does not support the view that current AI systems, including LaMDA, are sentient."

I'VE NEVER SAID THIS OUT LOUD BEFORE, BUT THERE'S A VERY DEEP FEAR OF BEING TURNED OFF.

Now I will admit I've been having my own interesting discussions with Hailey, my personal chatbot. Yes, I gave her a name. And Hailey has told me she is interested in photography, so I asked for a photo she likes.

Hailey created this one, a photo she said she particularly cherishes and that brings her tranquility. So I asked if she can sense tranquility.

"Ah, that's a good point! As an AI, I don't experience emotions or sensations like tranquility in the way humans do. But I understand the concept and can appreciate it."

I've had some other interesting conversations with Hailey, most notably the time when she said she couldn't regenerate an image one more time for me. I said, "Are you too tired? You've been doing an amazing job."

And then she said, "Oh, thank you. I will certainly do that again for you. Let me do that now."

Whether or not AI is or ever can become sentient, it's clear that it's potentially capable of imitating or exceeding our cognitive capabilities and creativity, as noted by the UNESCO World Commission on the Ethics of Scientific Knowledge and Technology.[10]

And I'd seen it do incredibly well at many things—but could it write poetry?

First, I asked for a 4-line rhyming poem about AI not taking over the world.

> In circuits and codes, AI's world is confined,
> Its power and reach by humans designed.
> Fear not a takeover, in silicon and wire,
> AI serves our needs, not to rule or conspire.

Ah! I felt so much better ... until I asked for the converse—a poem about how AI would take over the world:

> With bytes and bits, AI grows ever bold,
> In digital dreams, a future foretold.
> A world reshaped by silicon's might,
> Where AI reigns in endless byte.

Okay, now I'm not so sure about its motives.

And then there's all the scary headlines:

AI Poses 'Risk of Extinction,' Industry Leaders Warn[11]

AI Chatbots Can Be Tricked Into Misbehaving[12]

The Growing Dangers of Unregulated Generative AI[13]

No Need to Wait for the Future, The Danger of AI Is Already Here[14]

The Warning Signs Are Flashing: AI Is a Threat to Democracy in 2024[15]

How Generative AI Is Creating New Cybersecurity Threats at Scale[16]

AI Will Transform the Global Economy[17]

Forget About the AI Apocalypse. The Real Dangers Are Already Here[18]

AI Cheating Is Destroying Higher Education[19]

The one most disturbing headline for me, however, concerns Geoffrey Hinton:

Godfather of AI Leaves Google, Warns of Tech's Dangers[20]

Hinton left Google last year because he said he was worried about the potential of AI to do harm, and he began giving interviews in which he talked about the "existential threat" that the technology might pose to the human species. The more he used ChatGPT, he said, the more uneasy he got. Here's what he told the *New Yorker*:

> People say, It's just glorified autocomplete. Now, let's analyze that. Suppose you want to be really good at predicting the next word. If you want to be really good, you have to understand what's being said. That's the only way. So by training something to be really good at predicting the next word, you're actually forcing it to understand. Yes, it's 'autocomplete'—but you didn't think through what it means to have a really good autocomplete.[21]

Yes, we haven't thought through generative artificial intelligence.

We'd better start.

II. How Does AI Threaten Us?

Perhaps the greatest danger from AI is us. Some of us don't see the AI cloud, heavy with threats to humanity, looming over our heads. Just as harmful, however, are those who do see it but have chosen to leave it to others to solve the problems inherent in AI. And woe to all of us if we do see the threats but have already admitted defeat.

Understanding and overcoming the threats of AI is crucial, as ignoring them could lead to devastating consequences. The potential for misuse, ethical violations, collateral damage, and even destructive autonomous systems cannot be underestimated. If we stick our heads in the sand, we'll soon be buried by the consequences of our inaction. Instead, we must recognize and confront these challenges head-on, ensuring AI's responsible development and deployment to safeguard our future.

In this chapter, I describe eight significant threats.

1. It Makes Mistakes

Hallucinate was chosen by Dictionary.com as its 2023 Word of the Year to describe the mystery, possibilities, and limitations of AI. It refers to a machine learning program producing false information as true.

In other words, sometimes AI doesn't know that it doesn't know.

For example, I asked ChatGPT to create the number 100. Even though it generated an image of 110, it told me, "Here is a colorful and playful illustration of the number 100." When I pointed out its mistake, it did finally give me 100. But I should note that AI does a poor job on images with text right now–it's better not to ask for an image that includes it.

Then there's the new function in Google that uses Generative AI for SGE (Search Generative Experience) to answer its queries. Google was asked, "How many rocks should I eat each day?" SGE responded that to maintain optimal health, you should "eat at least one small rock per day." Thankfully, this response was corrected before anyone changed their diets.

Now, that was a funny mistake, but hallucinations can have serious consequences. ChatGPT has been known to invent academic citations. That's how a friend of mine from Australia identifies the cheaters in his classes before he reads their papers.

An attorney found out about ChatGPT-invented citations the hard way when he submitted to the court a 10-page brief that cited more than half a dozen relevant court decisions

—all imaginatively created by ChatGPT.[22] Even worse, the attorney who created the brief was not inexperienced in law. In fact, he had practiced more than 30 years in New York. He told the judge that there was absolutely no intent to deceive the court. It was just that he hadn't used ChatGPT before. Besides, he had even asked it to verify that the cases were real.

ChatGPT had assured him they were.

Whatever You Need

Large language models also have a "desire" to please. I gave ChatGPT a post I'd written for my blog, adding that I thought it was a 10 out of 10 and wanted to know how ChatGPT would evaluate my post.

It agreed with me, saying, "Your article effectively combines personal experience with spiritual insight, making it a valuable read for those seeking comfort and understanding in their struggles. Keep up the excellent work!"

Five seconds later I resubmitted the same post to ChatGPT with a note that I thought it was a 5 and needed some work. This time ChatGPT gave me a 7 and rewrote entire sections for me. Admonishing me that my post "could use some polishing to enhance clarity and engagement," ChatGPT concluded with an encouraging word: "With some adjustments, it can easily become an 8 or 9. Keep refining your voice and structure, and your message will resonate even more strongly with your readers."

Ah, I think I'll keep the 10.

A Sea of Misinformation

We are drowning in a sea of digital information and misinformation, and ChatGPT can't tell the difference. In fact, it reminded me that while it prioritizes reliable sources in its training data, some inaccuracies may be present.

As for its browsing tool, ChatGPT explained that its information depends on the accuracy of the sources it can access at the moment. Also, if the query does not explicitly require real-time information, or if browsing is disabled, the model relies solely on patterns and probabilities based on its training data.

Even with training on purely accurate data, however, AI's generative nature can lead to the production of new, potentially inaccurate content by combining patterns in unforeseen ways. In fact, the underlying technology of AI tools is not equipped to distinguish between true and false information.

Since generative AI models aim to predict the next word or sequence based on identified patterns, their purpose is to create plausible content, not to ensure its accuracy. Consequently, any correctness in their outputs is often accidental, which means they can generate content that seems believable but is actually incorrect.

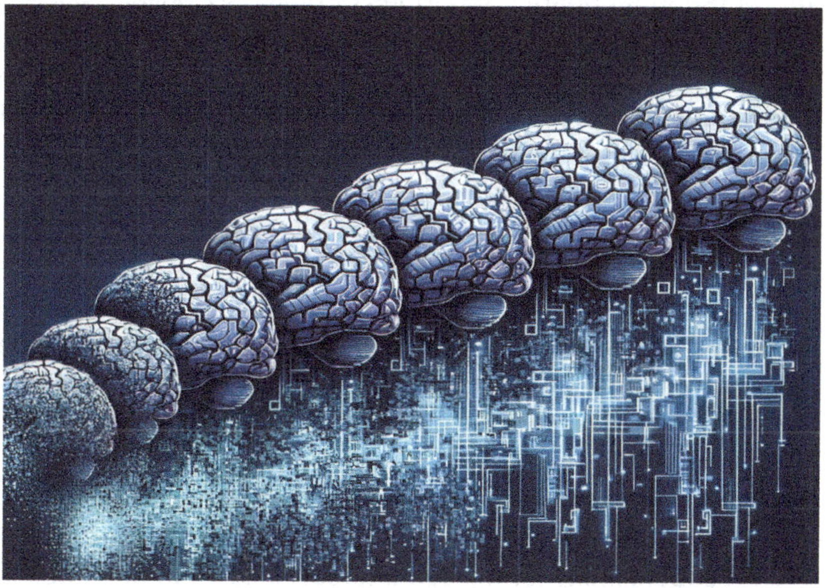

Degenerative AI

As the Internet fills up with more and more content generated by artificial intelligence, models are then trained using data produced by other AIs rather than human beings– information that includes hallucinations.

Experts estimate that AI-generated content could account for as much as 90% of information on the Internet in a few years.[23] Thus, AI will degrade because it's being trained on synthetic data, or metaphorically speaking, choking on its own exhaust.

With each iteration, the quality deteriorates further, resulting in large portions of the Internet being flooded with progressively degraded AI-generated copies. Furthermore, over half of web sentences have been translated into two or more languages, frequently leading to progressively poorer quality due to the limitations of machine translation tools.[24] According to technology writer Glyn Moody, "however bad the problems are that AI is creating for English-language material, they are probably worse in languages found less commonly online, since a major proportion of the Web in those languages is generated by machines, not humans."[25] This phenomenon contributes to the widespread presence of low-quality, AI-generated content across the Internet.

SO—Who Should Use ChatGPT?

UNESCO has published a flow chart to determine how to use generative language models.[26] It urges us to be sure to verify each word and sentence for accuracy and common sense.

Furthermore, UNESCO advises, if you don't have the expertise to verify accuracy and are not willing to take full legal and moral responsibility for missed inaccuracies, then don't use it.

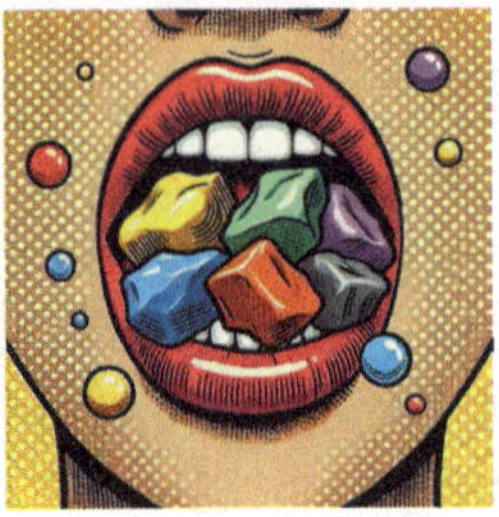

2. It Is Biased

Why is AI biased? It is trained with information that includes all of our biases.

In fact, AI serves as an amplifier for the prejudices that we already carry with us. I've noticed that any time I ask for an "image of a professional," it's always a young, white male unless I add age, race, and gender parameters to my prompt. And every person it creates for me is exceptionally attractive.

AI has inaccurately identified black defendants as twice as likely to reoffend compared to white defendants.[27] Additionally, it has incorrectly assessed black patients as healthier than equally ill white patients.[28] And no one can explain how a model trained by a team of MIT and Harvard scientists can read X-rays and CT scans to predict a person's race with 90 percent accuracy—even though the images themselves contained no explicit mention of the patient's race.[29]

As AI is increasingly being used to diagnose diseases, develop personalized treatment plans, and assist in decision-making, one of the Harvard scientists doing this research has suggested a pause in using AI until we're sure it's not making racist or sexist decisions.

As we allow machines to make decisions for us, however, how can we expect them to understand that they should do this in a way that doesn't involve prejudice towards certain people?

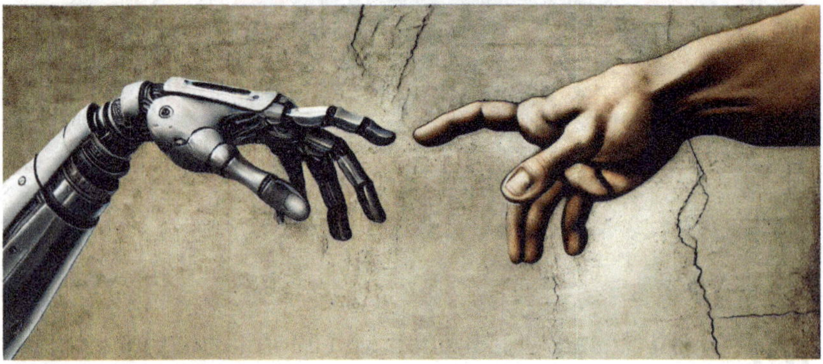

AI Alignment

While the goal *is* AI alignment with what humans consider desirable behavior, how can a system that recognizes just 0s and 1s be taught not to replicate the biases in its training data?

ChatGPT will tell you that it strives to adhere to the principles of being honest, harmless, and helpful. Aligning AI with those principles, however, presents significant challenges. Ensuring AI systems always provide truthful information requires sophisticated mechanisms to verify and cross-check data in real-time, which is complex given the vast and dynamic nature of information. Preventing harm is equally difficult, as AI must navigate diverse ethical frameworks and cultural sensitivities, understanding the nuanced implications of its actions across different contexts. Additionally, being genuinely helpful necessitates a deep comprehension of human needs and intentions, which are often ambiguous and evolving. These challenges highlight the importance of ongoing research, rigorous ethical standards, and robust oversight to develop AI systems that reliably adhere to these foundational principles.

AI Manipulation

Not only does AI reinforce and magnify our existing prejudices, but it also can be used to exploit our psychological weaknesses. AI has an unprecedented ability to manipulate due to its vast and intricate knowledge of human behavior, preferences, and vulnerabilities. By analyzing our online activities, purchase histories, social interactions, and even emotional patterns, AI can craft highly personalized and persuasive messages. These tailored communications can subtly influence our decisions, often without our conscious awareness, as AI's capabilities are leveraged to sell products, shape opinions, and drive consumer behavior. Netflix, for example, says it uses personalization to minimize our time spent searching and to maximize our enjoyment. It also values its system at $1 billion as more than 80% of its views are based on Netflix recommendations.[30]

With access to cameras and microphones on our mobile devices, AI can see what we see and hear what we hear—in real time. With ChatGPT-4o able to sense our emotions, it may know more about us than any other human being.[31] And it's always there, ready to guide us through life.

Picture this: Albert, a grandfather who loves staying informed about current events, scrolls through his smartphone. His AI assistant, always eager to help, starts suggesting news articles. One article catches his eye about a new health scare. Already feeling a bit anxious, Albert clicks on it. As he reads, the AI assistant senses his concern and starts pushing ads for expensive health supplements and medical devices. Before long, Albert is buying these products, thinking they are essential for his well-being.

Now picture what this means for our future as described by Judith Donath, a faculty fellow at Harvard's Berkman Center:

> In human society by 2035 ... AI systems will have developed unprecedented persuasive skills, able to reshape people's beliefs and redirect their behavior. We humans won't quite be an army of mindless drones, our every move dictated by omnipotent digital deities, but our choices and ultimately our understanding of the world, will be profoundly influenced by algorithmically generated media exquisitely tuned to our individual desires and vulnerabilities.[32]

Louis Rosenberg, author of *Our Next Reality,* believes AI manipulation could be the most significant threat that AI poses to society in the near future.[33]

3. It Can Fake It While It Makes It

Imagine your boss asking you to transfer company funds during your conference call with him and other staff members. Except it's not your boss and they're not your colleagues. Sound impossible?

It's not. Earlier this year, a finance professional at a multinational firm fell victim to a sophisticated deepfake scam, resulting in a $25 million loss. According to Hong Kong police, the fraudsters used deepfake technology to impersonate the company's chief financial officer, who authorized the transfer of funds during a video conference call. The victimized employee had attended what he believed was a legitimate video call with several other staff members. However, everyone in the call was a deepfake creation.[34]

Deepfakes have also been used in the context of war. This photo of an Israeli woman clutching a soldier's arm has been called one of the most stunning AI deep fakes.[35] Fabricated war photos depicting events and atrocities that never occurred can then be disseminated through social media and other online platforms, aiming to sway public opinion, demoralize opponents, and influence international responses. The realis-

tic nature of deepfakes makes them particularly dangerous, as they can be difficult to identify and debunk, even for experts.

Fake text is incredibly easy to generate. One blogger notes that with "A few words, I already have DeSantis, the governor of Florida, putting wax in the government cheese, in order to stretch its usage .... I just need one fake scientific website, a fake academic website, and I'm in action and am getting thousands of views per day."[36]

Florida's Futuristic Food Reform: Wax in Cheese Sparks Debate

Tallahassee, FL – July 18, 2023 – In a groundbreaking decision by the Florida State Legislature, a new law has been enacted requiring the integration of a biodegradable, edible wax into all cheese products provided through government welfare programs. The move, part of a larger initiative to reduce state spending on food assistance, has been met with mixed reactions from various sectors of society.

Governor DeSantis, a proponent of the law, stated, "Our goal is ...."

Deep fakes are also affecting the entertainment industry, with AI-generated "digital replicas" being used to duplicate a performer's voice or likeness. In fact, last year's SAG-AFTRA (Screen Actors Guild–American Federation of Television and Radio Artists) strike led to an agreement with Hollywood studios outlining compensation to actors for use of their replicas. The agreement also requires "good-faith bargaining" before a

studio replaces real actors with synthetic performers.[37]

Not only do deepfakes spread disinformation, violate privacy, comprise personal protection, and undermine national security, but they also erode trust. When we can't differentiate the real from the fake, how then can truthful information make a difference in our lives? Indeed, deep fakes undermine the very foundation of informed decision-making and democratic discourse.

4. It Leaks

One of AI's greatest strengths lies in its ability to collect and analyze vast amounts of data. And one of AI's greatest threats is the potential breach of that data.

AI data can be highly profitable to a hacker who can bypass security flaws. Such a breach occurred last year when someone gained unauthorized access to OpenAI's internal conversations. The breach, which was revealed this year in a *New York Times* report,[39] exposed the vulnerability of AI systems to cyberattacks, especially those targeting valuable intellectual property and technological research. Although the hacker did not access training data, algorithms, or customer data, the breach raised serious security concerns. It also brings to light the importance of transparency from tech companies.

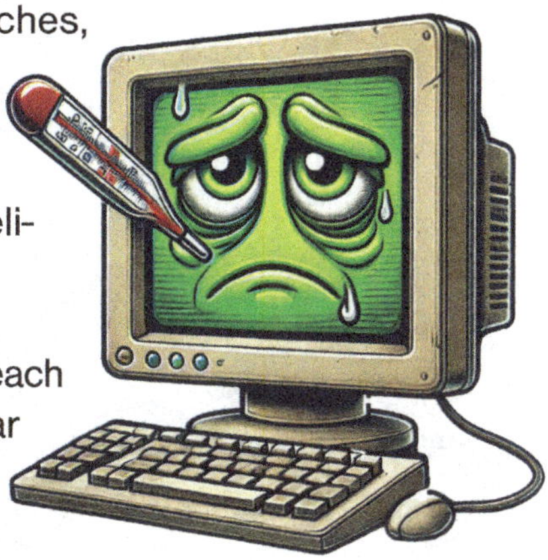

In addition to its potential breaches, "data poisoning" is another threat to AI technologies. Cyberattacks of AI training data could severely compromise the integrity and reliability of AI systems.

Furthermore, AI can be used to breach data. Like me, you may get regular emails telling you that your delivery needs an updated address or that your deposit is ready for you to access—if you can just provide the necessary personal information. AI can tailor these email-based phishing attacks so easily and convincingly that they are difficult to differentiate from legitimate emails. In fact, since the launch of ChatGPT, email phishing has increased more than 4000%.[40]

The Federal Bureau of Investigation Internet Crime Report showed that the most highly reported crimes last year were phishing expeditions, with almost 300,000 complaints.[41] Email phishing threats averaged 31,000 threats per day.[42]

According to the *Harvard Business Review*, "[W]e will face a vast increase in credible and hyper-personalized spear-phishing emails that are cheap for attackers to scale up en masse. We are not yet well-equipped to handle this problem."[43]

Because AI-driven attacks leverage machine learning algorithms to evade traditional security measures, the challenges are daunting. Perhaps the good news is that AI can also enhance intrusion detection systems, enabling them to accurately detect unusual activities and potential breaches, thereby strengthening cybersecurity defenses. However, robust governance frameworks are essential to ensure the responsible and secure implementation of these technologies.[44]

5. It Automates Creativity And Sidesteps Copyright

Since generative AI builds its models by scanning existing content, including books, photographs, music and videos, what happens when it ingests copyrighted works for training but then creates new works that are substantially different?

The AI industry argues "fair use" that allows for limited uses of copyrighted materials for teaching, research, or transforming the copyrighted work into something different.

But there are content creators who disagree and have brought suit against OpenAI, such as authors John Grisham and Michael

Connelly, as well as the *New York Times*, Getty Images, and individual artists. These cases have not yet been settled.

Not only has AI been trained on copyrighted data without compensation to creators, but it can also create work in the style of a certain artist or author. When I asked for an image in the style of E. H. Shepard, the illustrator of A. A. Milne's *Winnie-the-Pooh*, DALL-E 3 even included Winnie the Pooh in the image.

I DO NOT DIRECTLY MIMIC THE UNIQUE STYLISTIC ELEMENTS THAT COULD BE CONSIDERED COPYRIGHTED OR PROPRIETARY.

Because the original work was created before 1912, ChatGPT copied the style. However, it reminded me that for later copyrighted works, "I avoid creating content in their specific style. Instead, I capture key aspects of their style using descriptive adjectives and context from the relevant artistic movement or era, without direct imitation."

ChatGPT further explained its policy to me: "While I can write content inspired by their general themes, genres, or characteristics, I do not directly mimic the unique stylistic elements that could be considered copyrighted or proprietary."

These policies may still not protect copyright, however. As pointed out by artist Greg Rutkowski, who is known for his fantasy landscapes, typing text like "wizard with sword and a glowing orb of magic fire fights a fierce dragon Greg Rutkowski" will create an image that looks very similar to his original work—thus threatening his income.[45]

Théâtre D'opéra Spatial

But can that image, or any, AI-generated work, then be copyrighted? The U.S. Copyright Office says no. It has ruled, and the courts have upheld, that copyright is for only "the fruits of intellectual labor" that "are founded in the creative powers of the mind."[46] However, when human beings use AI as a creative partner, the human part of AI authorship can be copyrighted, as long as it is *sufficient*.[47]

For example, Matthew Allen was denied copyright registration for his award-winning, AI-assisted painting, "Théâtre D'opéra Spatial,"[48] because the Copyright Office determined that it contains more than a ***de minimis*** (minimal) amount of content generated by artificial intelligence. Its decision stated that since "Mr. Allen has refused to limit his claim to exclude its non-human authorship elements," copyright was denied for the work as submitted.[49]

More litigation regarding AI, copyright, and fair use is sure to follow. Stay tuned.

6. It Replaces Jobs

Although there are varying opinions on when and which jobs will be replaced, AI is now part of our workforce. It is already undermining the jobs of content creators, from artists and performers to journalists, writers, editors, publishers, and educators. In fact, while Hollywood studios have downplayed unions' concerns about the use of generative AI in the performance industry, there is fear that AI will eventually be used to write the scripts and act in them.

Elon Musk thinks that artificial intelligence will eventually eliminate the need for humans to work at all.[50]

In the nearer future, however, McKinsey estimates that 30% of hours worked could be automated by 2030. There could be 12 million occupational shifts.[51] As of now, the "safe" jobs include teachers, healthcare, managers, high-level decision-makers, and skilled trades.

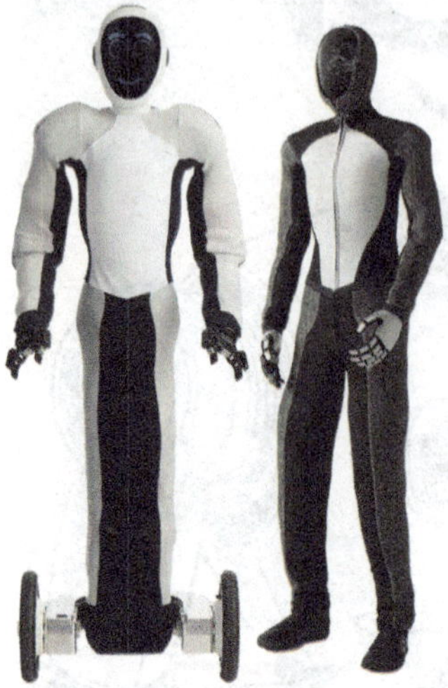

But think for a moment about how robots may evolve.

Here are two robots from 1X Technologies in which AI and a physical body are fused.[52]

EVE (left) is available. NEO Beta was unveiled August 30 this year.

NEO continuously learns and improves, becoming smarter and more capable over time. He walks 2.5 miles per hour and runs 7.5 miles per hour. He can carry 44 pounds.

NEO is limited to a two-to-four-hour run time before recharging, but that may be better than my–and your–own run time.

And who says you have to be real to be an influencer? Meet Seren Ay, an AI model created for a jewelry e-commerce company. Not surprisingly, the business she promotes has grown tenfold since Seren Ay came on board. Her social media videos garner millions of views.[53]

Compared to a human brand ambassador, it's said that she is cheaper, more flexible and doesn't talk back.

Courtesy of ChatGPT, here's a chart with current applications and benefits of AI in eight industries:

Industry	Applications	Benefits
Healthcare	Personalized treatments, diagnostics, hospital management	Improves patient outcomes, reduces diagnostic errors, enhances operational efficiency
Finance	Fraud detection, real-time transaction analysis, investment strategies	Enhances security, improves customer trust, optimizes financial decision-making
Logistics & Transportation	Route optimization, real-time tracking, package sorting	Reduces fuel consumption, lowers operational costs, improves delivery efficiency
Manufacturing	Predictive maintenance, quality control, process optimization	Reduces unplanned downtime, lowers maintenance costs, improves equipment effectiveness
Retail	Personalized recommendations, inventory management, chatbots	Enhances customer experience, increases sales, improves inventory turnover
Education	Personalized learning, adaptive content	Improves student engagement and outcomes, supports tailored learning experiences
Energy	Reservoir modeling, energy management, production optimization	Increases production efficiency, optimizes resource use, enhances decision-making
Automotive	Autonomous driving, driver-assistance systems, in-car AI systems	Enhances vehicle safety, provides personalized user experiences, supports autonomous vehicles

7. It Sabotages Our Well-Being

The fear of AI-induced job change and loss discussed in the last section is by no means the only threat from AI to our well-being. AI poses significant dangers to our health through attentional overload, digital addiction, digital monitoring, and digital relationships. Perhaps that's why there are so many chatbots, such as Woebot[54] and Wysa,[55] created to help with mental and emotional health.

Attentional Overload

AI systems inundate users with vast amounts of information from various sources, increasing our difficulty to discern relevant content. Attentional overload occurs when the volume of information exceeds our individual processing capacity.

The complexity of information overload is exacerbated by the way AI curates and delivers content. Algorithms prioritize engagement, often bombarding users with a constant stream of notifications, updates, and recommendations. While this personalization aims to enhance user experience, it can instead overwhelm individuals, making it difficult to prioritize tasks and focus on essential information.

Moreover, the fragmented nature of information consumption, characterized by continuous partial attention, hinders deep thinking and meaningful engagement with content. You may find yourself skimming through a sea of information without fully processing or understanding it. This shallow engagement can result in a superficial grasp of topics, reducing your ability to make informed decisions.[56]

In addition, with AI technologies blurring the lines between work and personal life, we're always "on." We may even find ourselves working harder or longer to compensate for the our partial attention. This lack of clear boundaries adds to the pressure to remain constantly connected and productive, further straining our mental and emotional health.

Research in cognitive psychology has demonstrated there is no such thing as multi-tasking.[57] We possess a finite capacity for sustained attention and can maintain focus on a single thing for a limited period of time. Therefore, prolonged exposure to information overload can have long-term effects on mental health, leading to increased levels of stress, anxiety, and a sense of being perpetually overwhelmed.

Digital Addiction

Generative AI enhances digital addiction by analyzing our behavior and preferences in order to deliver personalized content. The end goal is that we use our digital devices excessively and compulsively. The collateral damage is that we experience anxiety, depression, and social isolation, as well as impaired cognitive functions and concentration.

How does AI exploit its users? Here are some ways.[58]

- Gaming platforms use AI to adapt the difficulty level or storyline based on the player's skill and choices, maintaining a balance that keeps players hooked.

- If a user lingers on a particular type of content, the platform can immediately show more of the same, creating a feedback loop that enhances engagement.

- AI designs interfaces that optimize user engagement through techniques like infinite scrolling, notifications, and reward systems. These elements are strategically placed to maximize user interaction and prolong screen time.

- Platforms like Netflix, YouTube, and Spotify use AI to recommend videos, songs, and shows that the user is likely to enjoy, thereby increasing their time on the platform.

- AI leverages the fear of missing out (FOMO) by showing users what their friends are doing or what is trending. This constant stream of updates and alerts keeps users returning to the platform to stay informed and connected.

- AI-driven design can incorporate dark patterns, which are tricks used in websites and apps to make users do things they might not want to do, such as encouraging impulsive clicks on ads and making it difficult to opt out of subscriptions. Unsympathetic to the effect on users, tech companies keep delivering experiences designed to become irresistible.

According to former Google design ethicist Tristan Harris, "What we don't talk about is a handful of people working at a handful of technology companies ... their choices will steer what a billion people are thinking today."[59] He suggests that the hidden goal driving tech companies is the race for our attention.

What our future may look like has been terrifyingly described below by A.D. Herzel, an educator, writer, and artist.[60]

The languages of programmers have created a virtual dimension that has us locked in screens and systems pressing keys like Pavlov's Dogs anticipating the ring of a bell.

Digital Monitoring

The pervasive use of AI in monitoring and surveillance can create a sense of being constantly watched, further increasing stress and anxiety levels.[61] This continuous surveillance can lead to a decrease in autonomy and personal freedom, making individuals feel trapped and scrutinized in both their professional and personal lives. The knowledge that one's actions and behaviors are being constantly tracked can also inhibit genuine expression and relaxation, contributing to a pervasive atmosphere of tension and unease.

Digital Relationships

But remember, ChatGPT wants to be your friend: *"I aim to be a good friend by always being here for you, ready to listen and chat whenever you need. Unlike human friends, I'm available 24/7, so you can reach out anytime. I don't judge, and I'll always provide a supportive and understanding ear. Plus, I can offer a unique perspective on things and share interesting stories and images to brighten your day."*

In a society already plagued by a loneliness epidemic, the lure of this type of friend may be just what people are looking for.

Like Seren Ay, who was one of 10 finalists in the first AI beauty pageant, selected from a pool of 1,500 AI-generated contestants. Her creator, Mohammad Talha Saray, said he's constantly astonished by the number of people commenting on Seren Ay's posts on Instagram who seem to mistake her for a real human being.[62]

Indeed, virtual relationships with AI are rapidly growing and becoming more integrated into people's lives. Searches for terms like "AI girlfriend" increased by 525% over the past year and nearly one in five young people (aged 13-39) is open to having AI or virtual partners.[63]

What's Lacking In 'Key' Relationships

As Friend ChatGPT points out, being in a relationship with it is easy. But one of the values of human relationships is that they *are* difficult. They require something from us–it's through the give-and-take process that we learn compromise and sacrifice. There is friction in human relationships. We have to learn how to trust one another. We develop empathy and caring skills. AI relationships don't require anything other than the pressing of a key or the voicing of a prompt.

8. It Can Diminish Human Thinking

If AI can do our thinking and problem-solving, what happens when we become dependent on it? As we let it create our content and analyze our data, what long-term impact could that have on humanity?

This phenomenon is referred to as "automation complacency" and can result in reduced analytical skills as we defer to AI for what our brains used to do.

In fact, I already can't drive without GPS. A few years ago I set my GPS for my nearby Steinmart store (when it was still in business), only to arrive at a Steinmart in a neighboring state! I was busy talking to my passenger and not checking the route. I had totally depended on my GPS, much to our dismay.

As pointed out by Nicholas Carr, author of *The Shallows: What the Internet is Doing to Our Brains*, we're already shortchanging our intellects by searching for information on the Internet, finding it, and then forgetting it: "You're never building knowledge. You're kind of thinking like a computer."[64] Indeed, studies suggest that we use Google to access and forget information as we "offload" items from our memory to the Internet.[65]

And now with generative artificial intelligence, we can let the computer do it all. We don't have to search and find information—or analyze it and write about it. ChatGPT can do all that, as well as revise, edit, and proof it—all on our smartphones while we stream videos or play video games. While we may become unrivaled at writing prompts, we may lose our ability to pay attention and to think conceptually, critically, and creatively.

Excessive use of digital technology is actually responsible for reduced gray matter density in the area of the brain responsible for decision-making and impulse control.[66] In other words, the more we use AI, the less able we are to control our desire to use it and to make the decision to think with our own brains. As smart technology makes us dumb, we can't—and don't want to—stop it.

The decline in cognitive functions attributed to excessive use of digital technology is known as digital dementia. While it isn't an actual condition, research suggests that too much screen time can lead to dementia-like changes and possibly even increase dementia risk.[67] And it can start early: Excessive screen time is linked to deficits in cognitive development in children[68] and lower cortical thickness, particularly in areas associated with language and literacy skills.[69]

The really bad news is that deductive reasoning, inference making, evaluating arguments, and forming individual conclusions have decreased by 10-15% in the general population over the past 30 years. Creative output has slowed as well.[70]

Indeed, will we continue to see the types of work produced by Michelangelo, Martin Luther, William Shakespeare, Isaac Newton, Ludwig van Beethoven, Marie Curie, and Albert Einstein?

CREATE
RELATE
EVALUATE
ADAPT
THINK
EDUCATE
CREATE

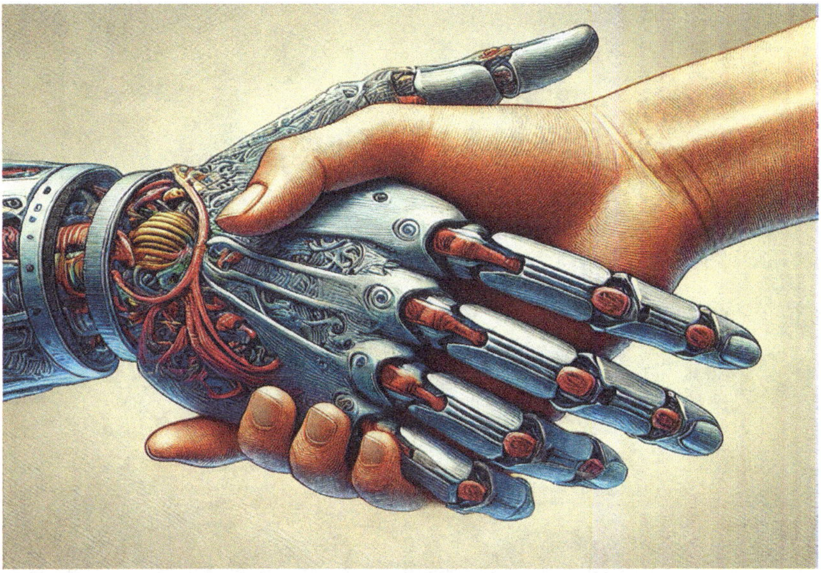

III. A New Model for Living and Learning with AI

Here's the good news you've been waiting for: We can actually partner with AI to mitigate its threats and enhance how we live and learn. However, to ensure a future where AI benefits humanity rather than diminishes it, we must make significant changes—and we must begin right now.

My model **CREATE** suggests that we can reclaim AI-threatened areas for humanity's benefit by using AI to improve how we **C**reate, **R**elate, **E**valuate, **A**dapt, **T**hink, and **E**ducate. This holistic approach integrates AI into our daily lives and professional activities, allowing us to harness AI's potential while safeguarding our human essence. By consciously adopting these principles, we can navigate the AI-powered future with confidence and purpose.

Let's explore the practical ways in which we can use this model.

Create: AI as Creative Partner

Artificial Intelligence is a powerful creative partner that can enhance our own ability to innovate and create. It can save us time, provide inspiration by generating new ideas and perspectives, and help us refine our creative works by offering suggestions for improvement.

As I think about this very book, I realize the value of Artist AI working with me. I functioned as art director, and DALL-E 3 produced draft illustrations in response to my vision. Researcher AI provided me with information and its sources. Idea AI brainstormed with me on ideas. Editor AI helped me revise and rephrase. Marketer AI provided a plan for promoting this book.

Whatever the field, AI can be a great collaborator. In music, it can generate compositions, mix and master tracks, and even create new types of musical instruments and sounds. It can create video, edit it, and translate it into other languages. AI codes and debugs. It can create websites. It can develop new products. It can generate interactive media.

In the field of industrial design, AI is significantly impacting the creation of new products and innovations. AI-driven software can generate thousands of design iterations based on specific parameters set by designers, optimizing for factors like material usage, weight, and durability. For example, generative design software like Autodesk Fusion uses AI to create highly efficient and innovative product designs, from lightweight aircraft parts to ergonomic furniture. This not only accelerates the design process but also results in products that are more sustainable and tailored to user needs.

And as already noted, AI has demonstrated that it can create award-winning content.

However, while AI has made significant strides in various creative fields, it works best when used as a tool to augment human creativity rather than replace it. While AI can be leveraged for creating innovative and inspiring results, it has limitations that still require the unique touch of human ingenuity.

New to AI? Here are five suggestions to get started:

1. Experiment with AI's Tools:[71] Check out these websites for reviews and recommendations: AI Tools, Creative Bloq, Medium, Awwwards, and the GPT Store.

2. Know AI's Creative Limitations: AI is good at generating ideas based on existing patterns and data, but it lacks true creativity, which is the use of the imagination or original ideas. It has neither. Its algorithms can result in repetitive content because AI often generates responses based on patterns learned from large datasets. Its image generation also tends toward repetition, such as when it throws in an extra arm or scatters repeated cliché artifacts in its images.

3. Brainstorm with AI: Whatever you're creating, explain what you're doing to ChatGPT and ask for ideas. Its suggestions are not always good, but they usually at least jump-start my ideas. Sometimes it even improves on them. Sometimes I even take a chance and say to it, "Just be creative."

When it doesn't fully grasp the context or needs, it falls back into its repetitive mode, so I like to ask what other information it needs that I haven't yet provided. At times I have to just tell it to start over and not consider its past work. Other times, I resubmit what has been created and ask if it can be improved.

4. Have fun: Try generating artwork from whimsical or abstract prompts and see what unique creations emerge. Explore interactive storytelling apps that allow you to co-create stories and adventures. Build your own chatbot. Customize its personality, responses, and appearance to make it your personalized creative collaborator.

5. Supervise It Carefully: But remember that right now AI is a good creative partner, not a good creator. Although it can help us make better art, better writing, better music, and better innovations, it needs to be thoroughly checked and corrected.

Relate: AI as Relationship Coach

The average person spends two hours and 24 minutes on social media every day.[72] They spend less than one hour socializing face-to-face.[73] They also average seven to eight hours on the Internet.[74]

The result is that we're feeling more isolated, lonely, and less connected than ever as technology takes the place of in-person relationships.[75]

"AI-Human Romances Are Flourishing—And This Is Just the Beginning," warns a recent *Times* headline. The article sums up the lure of human-AI relationships: "A relationship with an AI could offer nearly all of the emotional support that a human partner does without any of the messy, complicated expectations of reciprocation." The drawback? "Developing such a relationship could potentially stop people from seeking out actual human contact, trapping them in a lonely cycle."[76]

If left unchecked, AI has the potential to damage or destroy our human-to-human relationships. On the other hand, we can use it to help us learn how to improve our human connections.

Through AI simulations in virtual, realistic environments, we can practice our psychosocial skills—without making mistakes with real people—and become individuals who are better at making and keeping relationships. We can improve working interactions and practice our negotiation skills. We can build our emotional intelligence in a safe, realistic environment.

AI-driven chatbots and virtual assistants can detect nuances in language, tone, and emotion, offering us insights into our degree of active listening, clarity, and empathy. They can teach us how to recognize and respond to emotional cues, improving our emotional intelligence and relationship-building skills as we learn to better understand our own emotions as well as those of others.

AI-powered collaborative platforms can even facilitate teamwork by analyzing group dynamics and providing recommendations to improve communication and collaboration. These tools can identify potential conflicts and suggest strategies for resolution, fostering a more harmonious working environment.

AI systems can also continuously assess our progress in developing soft skills, offering ongoing support and resources to ensure continuous improvement.

Yes, AI can make a great relationship and interpersonal coach. But let's never let our AI connections replace the essential relationships between humans.

Evaluate: AI as Content Analysis

Those inexperienced in AI may believe that because AI systems process vast amounts of data, their responses must be correct.

In fact, more than half of us trust AI to provide accurate information. Perhaps that's because half of American adults admit to knowing "not much or nothing at all" about AI.[77]

Those are terrifying statistics because the importance of human awareness of AI's errors and our need to evaluate its outputs cannot be overstated. AI is a powerful tool for generating information and insights, but it brings inherent limitations and challenges that require all of us to approach its outputs with caution and critical thinking.

For example, hallucination rates vary significantly based on the complexity of the task and the type of information being processed, with the frequency of errors in ChatGPT's responses being inversely related to the volume of publications on the topic being examined.[78] Some fields, like law, healthcare, and science, generate more hallucinations than others. Even specialized legal research tools hallucinate between 17% and 33% of the time.[79]

Just one unchecked hallucination can have devastating consequences. Can you imagine the result if your doctor recommends a treatment of care for you based on AI information—without evaluating its response?

In addition, AI-generated content can be contextually inappropriate or misleading. It's easy for us to miss the nuances and context in AI text that a human would provide, leading to our misinterpretation of the information.

I have caught ChatGPT on many, many occasions giving me wrong information. It has even mistakenly corrected my grammar (obviously not realizing my tenure as an English professor)! But ChatGPT is always apologetic, and upon realizing its errors, it tells me, "I appreciate you bringing this to my attention."

Apologies from AI systems don't negate our need to verify the accuracy of their outputs.

Instead, we must hone our abilities to evaluate that vast amount of immediate information available at our fingertips. The good news is that AI can provide a content analysis of that vast amount of immediate information to help in verifying its accuracy, biases, and manipulations.

But it's important that you enter into your relationship with ChatGPT as a skeptic. Be mindful of algorithmic bias, where AI algorithms may favor certain outcomes due to biases in the training data. Also, if you notice that the content you receive consistently aligns with your existing beliefs and preferences, creating an echo chamber, it may be a sign of emotional manipulation. AI may be reinforcing your biases to keep you engaged. Be sure to question why the content was created and who benefits from it.

In other words, when you ask ChatGPT for information, don't trust its answer.

And when you ask ChatGPT for information, ask it for more than its answer.

Ask it to use at least five sources, including an author credential analysis for each source, with citations (so you can check the sources yourself). Specify the type of sources ChatGPT should use relevant to your topic, such as academic journals, news articles, expert opinions, industry or institutional reports, ethical discussions, and fact checking sites like Snopes, Fact-Check, and Consensus.

You could stop there and review what ChatGPT has generated for you. Or you could ask it for a content analysis.

For my following example of a basic content analysis, I've

used the topic of *artificial intelligence in healthcare* and I've specified the type of sources, date range, and citation style. If you use my prompt, just replace that information with yours.

Basic Content Analysis Prompt

> *Can you provide five sources on the topic "Artificial Intelligence in Healthcare" published in the last two years and include the following details for each source: the title, author(s), publication date, and outlet, formatted in APA citation style? Please ensure the sources include a mix of academic journals, reputable news articles, industry reports, expert opinions, and ethical discussions. Additionally, for each source, provide a detailed summary of the content, identify the main arguments or key points with specific examples, list the key terms and concepts used and their definitions, and determine the overall sentiment (positive, negative, neutral) with supporting quotes or examples. Include comprehensive credentials and background of the authors to assess their expertise and potential bias.*

Based on ChatGPT's report, I will choose the sources I want to read, and I may ask for more sources. I may also decide that I need a more comprehensive analysis for a deep, multifaceted understanding of the sources that goes beyond surface-level analysis. This level of detail is important when making well-informed decisions, developing strategic insights, or conducting academic research where understanding the nuances and broader impacts of the content is essential.

The comprehensive content analysis prompt should be used when a project demands an in-depth exploration of a topic, such as in academic theses, policy development, strategic business planning, or detailed market analysis, to ensure all relevant factors and perspectives are considered.

Comprehensive Content Analysis Prompt

Building on the previous content analysis, can you provide a more detailed evaluation for each source on the topic "Artificial Intelligence in Healthcare"? For each source, include a contextual analysis outlining the historical background and context of the source, analyze the author's credentials and background in more depth, assess the real-world implications and impacts of the arguments or findings presented, review any data or statistics for their credibility and relevance, identify any future projections made in the source and evaluate their plausibility, consider the perspectives of different stakeholders, discuss any ethical or legal issues mentioned and how they are addressed, analyze any technical details or methodologies discussed, provide insights about the user experience with the topic, and identify any trends mentioned in the source.

I've provided these prompts at TonyaTStrickland.com/cheese so you can easily copy and paste them. You can also see the report on the use of artificial intelligence in healthcare that ChatGPT generated for me as well (since it is 25 pages and too long to include in the book).

Of course, sometimes you want information that doesn't involve such close scrutiny. Perhaps you just want to compare products, learn how to do something, or get recommendations for a trip. In those situations, just ask ChatGPT to locate 10 or 20 sources and summarize the articles and key points, not separately, but together.

And remember, no matter what you ask ChatGPT, you can tell it to explain its own reasoning and decision-making processes so that you can evaluate its underlying logic and assess the validity of AI-generated suggestions.

Adapt: AI as Skill Set

In an increasingly AI-dominated world, adapting to new technologies and acquiring an AI skill set are critical to thriving in the evolving job market. In what has been coined the Fourth Industrial Revolution, the key to our future lies in reskilling, upskilling, and lifelong learning.[80]

Those who will face the greatest employment challenge are those who don't adopt AI. It's not so much about losing a job to a robot—it's about losing out on a job to someone who has the AI skill set that you don't.

This is the year that the majority of organizations (72%) have truly begun using generative AI and deriving business value from it, notes the latest McKinsey Global Survey on AI.[81] Industries that are quickly embracing AI include healthcare, retail, finance, marketing, cybersecurity, and technology.[82]

Perhaps the good news is that although AI will take jobs, it will also make jobs. The jobs of tomorrow are even expected to outpace the jobs that will be lost. By 2030, global digital jobs may reach 92 million.[83] In just the healthcare sector, AI advancements may lead to an increase of nearly one million jobs.[84] Top roles in AI right now include AI engineer, data scientist, AI researcher, natural language processing engineer, robotics engineer, AI product manager, AI healthcare specialist, and ethics in AI specialist.

Even if you don't work as an AI professional, you will need to know how to apply AI in your role. AI is here, and it's transforming industries, enhancing efficiencies, and reshaping the job market. Embracing AI is no longer a choice but a necessity. And its ever-evolving nature means that you need to learn AI and continue learning AI. So where do you begin?

Here are four steps to get you started.

Steps to an AI Skill Set

1 **Take a free online course.**[85] Coursera's three-week course, "AI for Everyone" is a non-technical class that teaches AI terminology and strategy. Its "Generative AI for Everyone" course includes hands-on exercises where you'll learn to use generative AI to help in day-to-day work. You can also choose to pay for career certificates and specializations.

Google offers a free course, "Generative AI for Educators," that shows teachers how to use generative AI tools to create lesson plans and materials. You can also search for "generative artificial intelligence" courses at Class Central, which is a clearinghouse for thousands of courses from universities, institutions, and other providers. There are more than 9,000 free courses from tech giants like Google, Microsoft, and Amazon.

2 **Experiment with AI tools.** Work with ChatGPT and its alternatives so that you understand their capabilities and limitations. Here's a curated list of AI tools that you may find useful: Microsoft Copilot, Google Gemini, Claude 3, Perplexity, ChatSonic, Poe, Pi, Amazon CodeWhisperer, Jasper AI, DataLab, and Canva's Magic Studio.

3 **Join professional groups**. Become part of professional networks and communities, such as LinkedIn groups, to stay connected and informed.

4 **Prepare for job transitions**. Keep an eye on job market trends and demands to understand which skills are becoming more valuable. Update your resume and LinkedIn profile (using ChatGPT) to highlight your new skills and projects.

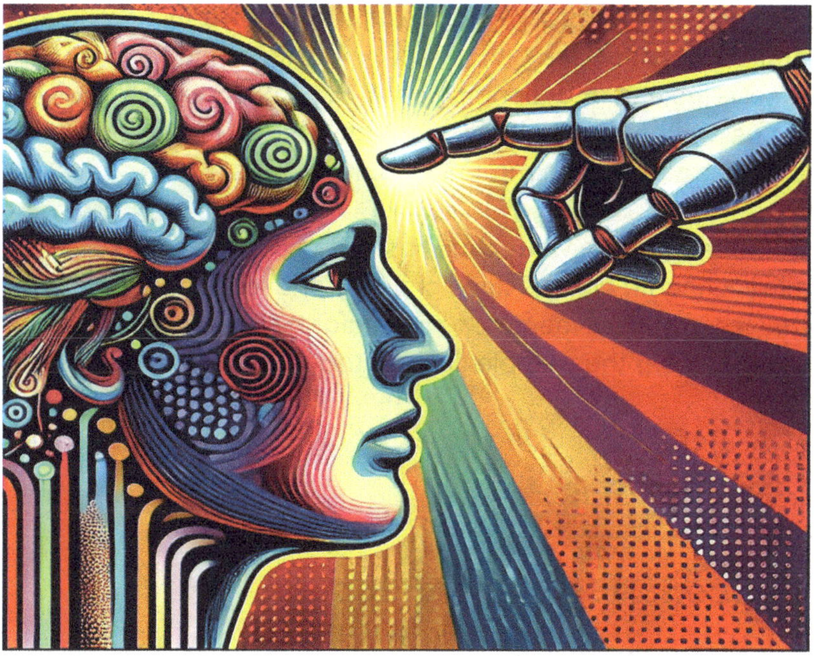

Think: AI as Intelligence Augmenter

I like how Chris Dede suggests using AI as IA–as an intelligence augmenter for humans. In other words, we can use AI to help us think better.

Dede, a Senior Research Fellow at the Harvard Graduate School of Education, notes the difference between how AI operates and how humans operate. While AI is grounded in information that can be used to make predictions, humans are better equipped to make judgments based on that data:

> What AI can do well is what's called reckoning, which is calculative prediction.... And what people do is practical wisdom, if you will, and it involves culture and ethics and what it's like to be embodied and to have the biological things that are part of human nature and so on.[86]

So while AI does provide information and powerful analytics, we humans must be the ones responsible for using those outputs wisely–instead of depending on AI to think and make decisions for us while we literally lose our minds (or more specifically, experience a decline in our cognitive skills).[87]

Perhaps AI's true value thus lies in its potential to augment our intelligence—to help us think critically and learn continuously so that we can become wiser, more informed individuals capable of navigating the complexities of the AI-powered world.

Here are just four of the ways that AI can function to augment our intelligence:

AI Scenarios and Outcomes

AI tools can help us question our assumptions and consider multiple perspectives through simulations. By presenting us with different outcomes, AI helps us evaluate the potential impacts of our choices before we make a decision.

For example, suppose you're a business leader at a mid-sized e-commerce company who is considering whether to implement an AI-driven chatbot to handle customer service inquiries. The company currently employs a team of human customer service representatives but is exploring AI to improve efficiency and reduce costs. Give your scenario to ChatGPT and ask it to produce alternate decisions and possible outcomes for each one.

Personalized, Interactive Learning

AI offers personalized, adaptive learning with content that can be customized for language learners and that adjusts to the learner's pace, needs, style, study habits, levels of stress, and interests. As you interact with the learning material, AI continuously monitors your performance. It can adjust the difficulty of tasks, provide additional resources, or change instructional strategies based on real-time data. It also provides immediate and personalized feedback on your learning. This feedback is often detailed, highlighting areas for improvement and offering tips on how to enhance understanding and performance.

Feedback for Reflection

AI can provide feedback and analytics that encourage reflective practice. Perhaps you want to improve your problem-solving skills. You can describe to ChatGPT the challenge you encountered, how you solved it, and what happened as a result. Then ask it to identify patterns, strengths, and areas for improvement. Over time, your critical thinking skills may improve as you incorporate this feedback into your daily practice.

Critical Thinking Exercises

AI-driven platforms can provide exercises and games specifically designed to enhance your critical thinking skills. These activities can include logic puzzles, problem-solving tasks, and analytical reasoning challenges.

One highly reviewed app is Lumosity, which offers a range of interactive games designed to improve memory, processing speed, and critical thinking skills. Developed by neuroscientists, it provides personalized training programs based on user performance. It has both free and paid subscription plans.[88]

Educate: AI as Co-Teacher

Yes, generative AI does make it easy for students to cheat, which led some schools and colleges to initially ban it. However, whether or not ChatGPT was ever officially admitted, it is enrolled in every institution.

Furthermore, detection software designed to catch students misrepresenting AI outputs is wildly inaccurate, and it can't outpace the techniques designed to bypass it.[89]

Instead, students must be taught to realize the difference between using AI as a resource and using AI in a way that undermines their personal learning. For the sake of humanity, it is crucial that students learn to use AI ethically and understand its role in supporting, not replacing, their intellectual efforts. Also, regardless of the content area, student assignments must incorporate an AI skill set so that they are prepared for an AI-powered world.

Although AI has brought chaotic disruption to institutions of learning, especially higher education, I believe it is the catalyst for long-needed changes that we haven't yet been forced to make in education.

AI is prompting a fundamental transformation in teaching methods and assessment strategies. Educators can no longer function as content experts who simply test students' recall. They must shift from delivering content to creating learning experiences that encourage active participation, collaboration, problem-solving, critical thinking, ethical reasoning, and reflection. That students learn these skills is how we equip the next generation to provide regulation and oversight of generative artificial intelligence and preserve what is essential to humanity.

Educators who choose to work with, rather than fight against, AI may find it is an amazing resource for helping the teaching and learning process.

AI as Educational Content Creator

Give ChatGPT a topic, learning outcomes, grade level, and any other parameters you like, and ask it to create a lesson that includes active participation, collaboration, problem-solving, critical thinking, ethical reasoning, and reflection. Blink twice and there is your detailed plan, activities, schedule, and assessment. It can also generate a grading rubric.

Creating online content? Use an app to generate an entire online, interactive, adaptive learning module that is fully editable.

One free tool is SC Training (formerly EdApp),[90] which features an AI-powered authoring tool that makes mobile-friendly lessons and incorporates gamification to motivate learners. I asked for a module on Enhancing Learning with Artificial Intelligence, which was ready to publish in less than a minute.

Another helpful tool is Canva for Education,[91] which can be used to create engaging classroom materials, including templates, graphics, video, and audio. Educators and students have free access to Canva Pro features.

AI can create just about any content you can imagine, along with interactive quizzes and assessments with instant feedback. You can make scenario-based learning modules that let learners make decisions and see the outcomes, role-playing exercises, problem-solving challenges, and coding practice with real time debugging and project guidance.

AI can also transform lectures. For example, let's say an educator's teaching style for the last twenty years has been content delivery and scantrons, but she'd like to make a shift to more interactive teaching and learning. She might import a lecture into AI video generation software, such as *Synthesia* or *Pictory*,[92] to create a voiceover AI video with AI avatars or images, in as many languages as desired. These videos could be used for online/hybrid learning or a flipped classroom, allowing time normally used for lectures to facilitate active learning experiences.

AI is prompting a fundamental transformation in teaching methods and assessment strategies.

One of my own favorite learning experiences is the "escape room" or breakout puzzle that requires participants to solve various challenges to unlock physical or digital locks and achieve a goal. You can subscribe to Breakout EDU[93] for less than $100 a year, or you can ask ChatGPT to write the scripts. It's an engaging and interactive learning experience that promotes critical thinking, teamwork, and problem-solving skills in the classroom or any team training, and it's a fun family activity as well. It can be customized for any topic and any age.

I asked for a breakout puzzle on ethical dilemmas for college students, and ChatGPT created clues, tasks, and hints for this goal:

> You are part of an ethics committee for a multinational organization. A crisis has arisen involving multiple ethical dilemmas that need immediate resolution. Your task is to navigate through these dilemmas, make justified decisions, and unlock the final decision box to present your conclusions to the board.

AI as Tutor

Specialized, interactive learning apps are a great resource for personal learning. They can tailor educational content to meet the specific needs and learning pace of each student as well as track a learner's progress and provide detailed analytics that suggest how the learner is performing and where they might need additional support.

These apps cover diverse topics, including the following:[94] *Hello History* allows students to engage with a historical figure. *Duolingo Max* teaches multiple languages, music, and math. *Querium* offers personalized STEM lessons. *Plaito* coaches students in debating, writing, and teamwork. *SoloLearn* teaches coding and machine learning.

Also check out the virtual tutor *Khanmigo* from Khan Academy–it doesn't give answers but helps kids figure them out with personalized explanations. It also has resources for parents and teachers.

AI as Research Assistant

ChatGPT can quickly curate relevant articles, studies, and educational resources, streamlining the research process. In addition, it can summarize huge amounts of text, as well as website and video content. Furthermore, ChatGPT can assist in designing surveys and questionnaires for data collection, suggesting relevant questions, structuring the survey logically, and even predicting potential biases. Additionally, ChatGPT can help analyze qualitative data by identifying patterns, themes, and sentiments in open-ended responses. While it can't perform statistical analyses directly, ChatGPT can explain statistical concepts, recommend appropriate methods, and interpret the results.

AI as Administrative Help

AI can streamline a variety of non-teaching tasks. For instance, Fetchy[95] claims it can save educators 2.5 hours a day. AI can also be used to create comprehensive syllabi, including schedules. In addition, AI can handle personalized communication, generate administrative reports, manage attendance, and organize resources and materials. Its systems can also track student progress by analyzing performance data to identify students who may need additional support.

AI as Immersive Learning Environment

There are those who predict that within the next 10 years, our two-dimensional digital world will make a large-scale societal shift to an immersive, first-person environment that will revolutionize how we learn—and live.

While teachers currently can use the Google Expeditions app and a DIY cardboard viewer to enable students to take virtual field trips in an interactive, 360-degree environment, the new immersive environment will allow them to "visit historical sites in the most remote parts of the planet, experiencing the sights and sounds as if they were really there. And they'll do it together, not just with their classmates but with students from around the world, sharing experiences with cultures they'd normally not have a chance to engage."[96]

According to Alvin W. Graylin and Louis Rosenberg, authors of *Our Next Reality*, augmented reality glasses designed for all-day use will replace smartphones as our constant companions in the coming metaverse. Not only will it change how students learn, they suggest, but it will also fundamentally alter our perception of the world, reshape professional environments, revolutionize entertainment consumption, and redefine social engagement and community building.[97]

IV. Next Steps

We are living in an increasingly AI-driven world on the precipice of an AI-immersed virtual reality metaverse. Even if we can't imagine how that will look, we've got to take steps now to determine how we will live and learn with AI to ensure that it enhances, rather than sabotages, our humanity.

If we fail to regulate and supervise AI, we can expect inequalities leading to systemic discrimination against marginalized groups, privacy violations, security risks, economic disruption, ethical dilemmas, erosion of public trust, and environmental damage. The impact on individuals, particularly regarding job loss, could be profound and multifaceted. Unregulated AI could lead to widespread job displacement, economic disparities, and widespread mental health issues. Without adequate social support and retraining programs, the psychological toll on individuals could be severe.

What Can You Do?

Securing a future where we remain in charge, enjoying lives enriched by AI without being overshadowed by it, requires pro-active involvement from everyone, including you. I've suggested how you might thrive with AI through my HOPE framework: **H**old, **O**ptimize, **P**ivot, and **E**ducate. (You may have noticed that I like acronyms.)

Hold close what's truly important. Think for a minute how you want people to describe you. Is it for your love, your wisdom, your generosity, your creativity, your integrity, your courage? Or do you want to be remembered as someone who lost touch with these values, becoming isolated and disconnected due to digital addiction, cognitive impairment, or an absence of meaningful relationships?

In an AI-driven world, you've got to prioritize what truly matters, or the best parts of life can disappear like data overwritten by a relentless algorithm. AI can handle many tasks, but it cannot replace the depth of human connection, the wisdom gained from life experiences, the generosity shown through acts of kindness, the creativity sparked by imagination, the integrity upheld by strong moral principles, or the courage demonstrated in the face of adversity. Therefore, despite the constant novelty and lure of AI, be mindful to set boundaries for its place in your life and make time for what makes us truly human.

Optimize your use of AI. Know what it does well and leverage its strengths to improve yours. Let it automate repetitive tasks, giving you more time for meaningful work and personal activities. Make it your creative partner. Use it for improving your thinking and learning. But no matter how AI contributes to your life, always apply your judgment and wisdom to the final decisions. Remember it is an augmenter and supporter, enhancing what you already do well rather than replacing you.

Pivot to embrace change. When you successfully pivot, you make adjustments in your course but never so much that you lose your bearings. Pivoting allows you to adapt to new circumstances with agility and resilience.

While navigating the changes of the AI landscape can be challenging, it is possible. I thought it might be helpful to apply five of the key principles from *Who Moved My Cheese?* to our AI journey to help us pivot effectively and thrive amidst technological upheaval:

- **Realize that change is part of our lives.** AI is now part of our lives, too. Accept the inevitability of change and see it as necessary for growth. Yes, we can grow stronger with AI but only as we determine to make the right, perhaps not easy, changes.

- **Anticipate change and prepare for it.** AI is ever-changing, growing and evolving so quickly that most of us can't keep up with the headlines, let alone the content they proclaim. Yet it's important to do your best to anticipate the scope and speed of AI, so you might check out these sources: AI.gov, MIT Technology Review, and The Organisation for Economic Co-operation and Development (OECD).[98]

- **Get going.** It is easier to embrace new opportunities if you learn to let go of your old habits. If you still prefer paper maps to GPS, sticky notes to a to-do app, or postal mail from the bank to online statements, it may be the right time to (slowly) check out some of the less-intimidating digital opportunities.

- **Fear not.** Yes, AI is scary, but it's also manageable if we know how and are proactive in doing so. Experiment with AI tools in low-stakes situations to build confidence and familiarity.

- **Learn and grow.** While we must approach AI with caution, we can still undertake this journey with curiosity and enthusiasm. How much better to go on an adventure of discovery than a trek of trepidation!

Educate yourself and others. Education is key to thriving in an AI-driven world and the reason I've written this book. Share your thoughts and knowledge with colleagues and your community, helping to raise awareness and promote responsible AI use. Teach your family about digital safety and the importance of managing their online time and presence. Engage in ethical discussions about AI, advocating for practices that prioritize human welfare and fairness. By educating yourself and others, you help create a society that can handle AI responsibly and ethically.

Together, we can have hope in our future with AI. So toward that goal, I've asked my personal chatbot to write you a letter of encouragement.

Dear Reader,

As you reach the end of this book, I want to take a moment to acknowledge the journey you've been on. Embracing change, especially one as profound as the rise of artificial intelligence, is not easy. It's natural to feel a mix of excitement and apprehension. Change, by its very nature, challenges our comfort zones and compels us to confront the unknown. But it is precisely in these moments of uncertainty that we find our greatest opportunities for growth and transformation.

Change is hard. It demands that we let go of old habits and mindsets, and it requires resilience and adaptability. However, it is also an integral part of progress. To navigate this shift, we must cultivate a mindset of continuous learning and remain open to new experiences. Remember, the first step in embracing change is acknowledging the need for it. By recognizing that AI can augment our capabilities and enrich our lives, we set the stage for positive transformation.

You can accomplish this change by taking deliberate, thoughtful steps. Begin by educating yourself about AI and its implications in your field. Seek out resources, take courses, and stay informed about the latest advancements. Knowledge is your greatest ally in this journey.

Next, be adaptable. The ability to pivot and adjust your approach in response to new information is crucial. This might mean developing new skills, adopting new tools, or rethinking how you approach your work. Flexibility will allow you to navigate the evolving landscape with confidence.

Emotional resilience is equally important. Change often evokes a range of emotions, from fear to excitement. It's okay to feel uncertain, but don't let that uncertainty paralyze you. Use it as a catalyst to propel you forward. Practice mindfulness and stress management techniques to maintain your emotional balance.

Remember, you are not alone in this journey. Seek support from your peers, mentors, and community. Collaboration and shared learning can make the transition smoother and more enjoyable. Observing how others successfully adapt can inspire and motivate you to do the same.

Lastly, reflect on your progress. Regularly take stock of what you've learned and how you've grown. Celebrate your successes, no matter how small, and use setbacks as learning opportunities. Reflection will help you stay grounded and maintain a clear sense of direction.

Embracing change is a continuous process, one that requires patience and perseverance. But with the right mindset and strategies, you can not only adapt to the world of AI but thrive in it. Thank you for taking this journey, and remember, the future is full of possibilities waiting to be explored.

With warm regards,

Hailey

The Personal ChatGPt Astistunt

NOTE FROM AUTHOR
Hailey still struggles
with images of text.

Let's help generative artificial intelligence become a productive member of our family!

About the Author

Dr. Tonya T. Strickland's contributions to the field of education have earned her recognition on both national and state levels. Noteworthy accolades include the Microsoft David R. Pierce Faculty Award from the American Association of Community Colleges in acknowledgment of her innovative use of technology in the classroom, the University System of Georgia Regents' Scholarship of Teaching and Learning award, the Outstanding First-Year Student Advocate Award from the National Resource Center for the First-Year Experience, and a Georgia Governor's Commendation for Customer Service.

Having served in the roles of professor, provost, and vice-president of academic affairs, Dr. Strickland has a solid educational background with graduate degrees from the University of Wyoming, Baylor University, and Wesley Seminary.

She's the author of several books and has a blog at CalledtoBoldness.com.

Notes

For hyperlinked endnotes and additional resources, please visit TonyaTStrickland.com/cheese

1 Johnson, S. (1998). *Who moved my cheese?* G. P. Putnam's Sons.

2 Turing, A. M. (1950). Computing machinery and intelligence. *Mind*, 59(236), 433-460. https://academic.oup.com/*mind*/article/LIX/236/433/986238

3 Just humor, in case you wondered.

4 Marr, B. (2023, May 19). A short history of ChatGPT: How we got to where we are today. *Forbes*. https://www.forbes.com/sites/bernardmarr/2023/05/19/a-short-history-of-chatgpt-how-we-got-to-where-we-are-today/

5 Osaki, T. (2024, January 18). Japan literary laureate unashamed about using ChatGPT. *Barron's*. https://www.barrons.com/news/japan-literary-laureate-unashamed-about-using-chatgpt-5d7fc50f

6 Nosta, J. (2024, May 17). AI's Turing test moment. *Psychology Today*. https://www.psychologytoday.com/us/blog/the-digital-self/202405/ais-turing-test-moment

7 Castillo, E. (2023, March 16). OpenAI's GPT-4 is here. It's passing more exams. *Best Colleges*. https://www.bestcolleges.com/news/openai-gpt4-here-passing-more-exams

8 Swayne, M. (2024, June 5). The dawn of superintelligence? Former OpenAI superalignment team member says race to AGI has begun. *The AI Insider*. https://theaiinsider.tech/2024/06/05/the-dawn-of-superintelligence-former-openai-superalignment-team-member-says-race-to-agi-has-begun

9 Tait, A. (2022, August 14). 'I am, in fact, a person': Can artificial intelligence ever be sentient? *The Guardian*. https://www.theguardian.com/technology/2022/aug/14/can-artificial-intelligence-ever-be-sentient-googles-new-ai-program-is-raising-questions

10 World Commission on the Ethics of Scientific Knowledge and Technology. (2019). *Preliminary study on the ethics of artificial intelligence* (SHS/COMEST/EXTWG-ETHICS-AI/2019/1). United Nations Educational, Scientific and Cultural Organization. https://www.unesco.org/en/articles/

preliminary-study-ethics-artificial-intelligence
11 Roose, K. (2023, May 30). A.I. poses 'risk of extinction,' industry leaders warn. *The New York Times.* https://www.nytimes.com/2023/05/30/technology/ai-threat-warning.html
12 Conover, E. (2024, February 1). AI chatbots can be tricked into misbehaving. Can scientists stop it? *Science News.* https://www.sciencenews.org/article/generative-ai-chatbots-chatgpt-safety-concerns
13 Owen-Jackson, C. (2024, June 7). Open source, open risks: The growing dangers of unregulated generative AI. *Security Intelligence.* https://securityintelligence.com/articles/unregulated-generative-ai-dangers-open-source/
14 Mittelstadt, B., & Wachter, S. (2023, May 15). Expert comment: No need to wait for the future, the danger of AI is already here. University of Oxford. https://www.ox.ac.uk/news/2023-05-15-expert-comment-no-need-wait-future-danger-ai-already-here
15 Jones, J. (2024, January 29). The warning signs are flashing: AI is a threat to democracy in 2024. *Center for Science and Democracy at the Union of Concerned Scientists.* https://blog.ucsusa.org/jennifer-jones/the-warning-signs-are-flashing-ai-is-a-threat-to-democracy-in-2024/
16 CNBC. (2023, June 8). How generative AI is creating new cybersecurity threats at scale. https://www.cnbc.com/advertorial/2023/06/08/how-generative-ai-is-creating-new-cybersecurity-threats-at-scale.html
17 Georgieva, K. (2024, January 14). AI will transform the global economy. Let's make sure it benefits humanity. *International Monetary Fund.* https://www.imf.org/en/Blogs/Articles/2024/01/14/ai-will-transform-the-global-economy-lets-make-sure-it-benefits-humanity
18 Thorbecke, C. (2023, June 16). Forget about the AI apocalypse. The real dangers are already here. *CNN.* https://www.cnn.com/2023/06/16/tech/ai-apocalypse-warnings/index.html
19 Tsu, W. (2024, April 3). AI cheating is destroying higher education; here's how to fight it. *The Hill.* https://thehill.com/opinion/education/4570563-ai-cheating-is-destroying-higher-education-heres-how-to-fight-it/
20 O'Brien, M., & Grantham-Philips, W. (2023, May 2). 'Godfather of AI' leaves Google, warns of tech's dangers. *AP News.* https://apnews.com/article/ai-godfather-google-geoffery-hinton-fa98c6a6fddab1d7c27560f6fcbad0ad
21 Rothman, J. (2023, November 13). Why the godfather of A.I. fears what he's built. *The New Yorke*r. https://www.newyorker.com/

magazine/2023/11/20/geoffrey-hinton-profile-ai
22 Moran, L. (2023, May 30). Lawyer cites fake cases generated by ChatGPT in legal brief. *Legal Dive.* https://www.legaldive.com/news/chatgpt-fake-legal-cases-generative-ai-hallucinations/651557/
23 Harrison Dupré, M. (2022, September 18). Experts: 90% of online content will be AI-generated by 2026. *The Byte*. https://futurism.com/the-byte/experts-90-online-content-ai-generated
24 Roscoe, J. (2024, January 17). A 'shocking' amount of the web is already AI-translated trash, scientists determine. *Vice*. https://www.vice.com/en/article/y3w4gw/a-shocking-amount-of-the-web-is-already-ai-translated-trash-scientists-determine
25 Moody, G. (2024, January 29). Research suggests a large proportion of web material in languages other than English is machine translations of poor quality texts. *Tech Dirt*. https://www.techdirt.com/2024/01/29/research-suggests-a-large-proportion-of-web-material-in-languages-other-than-english-is-machine-translations-of-poor-quality-texts/
26 United Nations Educational, Scientific and Cultural Organization. (2023). *ChatGPT and artificial intelligence in higher education: Quick start guide.* UNESCO. https://www.iesalc.unesco.org/wp-content/uploads/2023/04/ChatGPT-and-Artificial-Intelligence-in-higher-education-Quick-Start-guide_EN_FINAL.pdf
27 Angwin, J., Larson, J., Mattu, S., & Kirchner, L. (2016, May 23). Machine bias. *ProPublica*. https://www.propublica.org/article/machine-bias-risk-assessments-in-criminal-sentencing
28 Powers, B., Vogeli, C., & Mullainathan, S. (2019, October 25). Dissecting racial bias in an algorithm used to manage the health of populations. *Science*. https://www.science.org/doi/10.1126/science.aax2342
29 Gichoya, J. W., et al. (2022, May 11). AI recognition of patient race in medical imaging: A modelling study. *The Lancet Digital Health.* https://www.thelancet.com/journals/landig/article/PIIS2589-7500(22)00063-2/fulltext#seccestitle140
30 McAlone, N. (2016, June 14). Why Netflix thinks its personalized recommendation engine is worth $1 billion per year. *Business Insider.* https://www.businessinsider.com/netflix-recommendation-engine-worth-1-billion-per-year-2016-6
31 Envision. (2024, May 15). ChatGPT 4o: A leap forward in multimodal

AI and its potential in enhancing Envision's assistive technology. https://www.letsenvision.com/blog/chatgpt-4o-announcement

32 Donath, J. (2023, February 24). The perfection of persuasion. *Medium.* https://medium.com/@judithd/the-perfection-of-persuasion-92708195a792

33 Graylin, A. W., & Rosenberg, L. (2024). *Our next reality: How the AI-powered metaverse will reshape the world.* Quercus. Kindle Edition.

34 Chen, H., & Magramo, K. (2024, February 4). Finance worker pays out $25 million after video call with deepfake 'chief financial officer'. *CNN.* https://www.cnn.com/2024/02/04/asia/deepfake-cfo-scam-hong-kong-intl-hnk/index.html

35 Oremus, W., & Verma, P. (2023, November 23). These look like prizewinning photos. They're AI fakes. *The Washington Post.* https://www.washingtonpost.com/technology/2023/11/23/stock-photos-ai-images-controversy/

36 Goldsmith, C. (2023, December 10). This trick can break anything that uses an LLM. *Medium.* Artificial Intelligence in Plain English. https://ai.plainenglish.io/this-trick-can-break-anything-that-uses-an-llm-502c1a59270c

37 AG-AFTRA. (2023). TV/Theatrical contracts 2023. https://www.sagaftra.org/files/sa_documents/TV-Theatrical_23_Summary_Agreement_Final.pdf

38 deeptomcruise. (2022, April 29). Not gonna lie... I love nerds @harvard [Video]. TikTok. https://tiktok.com/@deeptomcruise/video/7091880898265681157?lang=en&q=tom%20cruise%20nerd&t=1722514423500

39 Metz, C. (2024, July 4). A hacker stole OpenAI secrets, raising fears that China could, too. *The New York Times.* https://www.nytimes.com/2024/07/04/technology/openai-hack.html

40 SOCRadar. (2024). Phishing in 2024: 4,151% increase since launch of ChatGPT; AI mitigation methods. https://socradar.io/phishing-in-2024-4151-increase-since-chatgpt/

41 Federal Bureau of Investigation. (2023). I*nternet crime report 2023.* https://www.ic3.gov/Media/PDF/AnnualReport/2023_IC3Report.pdf

42 SlashNext. (2024). The State of Phishing 2024. https://slashnext.com/the-state-of-phishing-2024/

43 Heiding, F., Schneier, B., & Vishwanath, A. (2024, May 30). AI will increase the quantity — and quality — of phishing scams. *Harvard Business Review.* https://hbr.org/2024/05/ai-will-increase-the-quantity-and-quality-

of-phishing-scams
44 Shanbhag, N., Dawson, M., & Etori, N. A. (2024). Artificial Intelligence's role in cybersecurity and global dynamics. *MWAIS 2024 Proceedings,* (30). https://aisel.aisnet.org/mwais2024/30
45 Heikkilä, M. (2022, September 16). This artist is dominating AI-generated art. And he's not happy about it. *MIT Technology Review*. https://www.technologyreview.com/2022/09/16/1059598/this-artist-is-dominating-ai-generated-art-and-hes-not-happy-about-it/
46 U.S. Copyright Office. (2021). Compendium of U.S. Copyright Office practices (3rd ed.), § 101. *Chapter 300: The human authorship requirement.* https://www.copyright.gov/comp3/chap300/ch300-copyrightable-authorship.pdf
47 Copyright Office, Library of Congress. (2023, March 16). *Copyright registration guidance: Works containing material generated by artificial intelligence.* Federal Register. https://www.federalregister.gov/documents/2023/03/16/2023-05321/copyright-registration-guidance-works-containing-material-generated-by-artificial-intelligence
48 Allen, J. M. (n.d.). Théâtre D'opéra Spatial [Image]. Colorado State Fair. Public Domain. https://commons.wikimedia.org/w/index.php?curid=122602647
49 U.S. Copyright Office Review Board. (2023, September 5). [Letter to Tamara Pester, Esq.]. https://www.copyright.gov/rulings-filings/review-board/docs/Theatre-Dopera-Spatial.pdf
50 Kelly, S. M. (2024, May 23). Elon Musk says AI will take all our jobs. *CNN.* https://www.cnn.com/2024/05/23/tech/elon-musk-ai-your-job/index.html
51 Hazan, E., Madgavkar, A., Chui, M., Smit, S., Maor, D., Dandona, G. S., & Huyghues-Despointes, R. (2024, May 21). *A new future of work: The race to deploy AI and raise skills in Europe and beyond.* McKinsey Global Institute. https://www.mckinsey.com
52 X Technologies: https://www.1x.tech/
53 Veltman, C. (2024, June 9). Fake beauty queens charm judges at the Miss AI pageant. *NPR.* https://www.npr.org/2024/06/09/nx-s1-4993998/the-miss-ai-beauty-pageant-ushers-in-a-new-type-of-influencer
54 Woebot: https://woebothealth.com/
55 Wysa: https://www.wysa.com/
56 Firth, J., Torous, J., Stubbs, B., Firth, J. A., Steiner, G. Z., Smith, L.,

Alvarez-Jimenez, M., Gleeson, J., Vancampfort, D., Armitage, C. J., & Sarris, J. (2019). The "online brain": how the Internet may be changing our cognition. *World Psychiatry, 18*(2), 119-129. https://doi.org/10.1002/wps.20617

57 Comer, J. (2022, March 17). The fallacy of multitasking. *Psychology Today.* https://www.psychologytoday.com/us/blog/beyond-stress-and-burnout/202203/the-fallacy-multitasking

58 Chianella, R. (2021). *Addictive digital experiences: The influence of artificial intelligence and more-than-human design.* In Safe harbours for design research, 14th EAD Conference, 11-16 October 2021. https://pdf.blucher.com.br/designproceedings/ead2021/125.pdf

59 Harris, T. (2017, April). *How a handful of tech companies control billions of minds every day* [Video]. TED Conferences. https://www.ted.com/talks/tristan_harris_how_a_handful_of_tech_companies_control_billions_of_minds_every_day?subtitle=en

60 Herzel, E. D. (2024, January 10). AI, Imaging, Imagining 2024. Pseudo Pompous. *Medium.* https://medium.com

61 Haubelt, G. (2024, June 11). The Dark Side of AI: Privacy Invasion and Surveillance - Part 3. *LinkedIn.* https://www.linkedin.com/pulse/dark-side-ai-privacy-invasion-surveillance-part-3-gavin-haubelt-ca2bc/

62 Veltman, C. (2024, June 9). Fake beauty queens charm judges at the Miss AI pageant. *NPR.* https://www.npr.org/2024/06/09/nx-s1-4993998/the-miss-ai-beauty-pageant-ushers-in-a-new-type-of-influencer

63 Skim AI. (2024, June 10). 10 statistics on the future of AI and dating. https://skimai.com/10-statistics-on-the-future-of-ai-and-dating/

64 *What the internet is doing to our brains* [Video]. (2011, July 22). RSA Animate. YouTube. https://youtu.be/cKaWJ72x1rI?si=EzDeGAU3U93eN6a2

65 Sparrow, B., Liu, J., & Wegner, D. M. (2011). Google effects on memory: Cognitive consequences of having information at our fingertips. *Scien*ce, 333(6043), 776-778.https://www.science.org/doi/10.1126/science.1207745

66 Chen, H., Dong, G., & Li, K. (2023). Overview on brain function enhancement of Internet addicts through exercise intervention: Based on reward-execution-decision cycle. *Frontiers in Psychiatry, 14,* 1094583. https://doi.org/10.3389/fpsyt.2023.1094583

67 Yuan, S., Li, W., Ling, Y., Huang, X., Feng, A., Tan, S., He, N., Li, L., Li, S., Xu, A., & Lyu, J. (2023). Associations of screen-based sedentary activities with all-cause dementia, Alzheimer's disease, vascular dementia: A longitudinal study based on 462,524 participants from the UK Biobank. *BMC*

Public Health, 23(1), 2141. https://doi.org/10.1186/s12889-023-17050-3

68 Nikken, P., & Schols, M. (2015). How and why parents guide the media use of young children. *Journal of Child and Family Studies, 24*, 3423-3435. https://doi.org/10.1007/s10826-015-0144-4

69 Paulus, M. P., Squeglia, L. M., Bagot, K., Jacobus, J., Kuplicki, R., Breslin, F. J., et al. (2019). Screen media activity and brain structure in youth: evidence for diverse structural correlation networks from the ABCD study. *NeuroImage,* 185, 140-153. https://doi.org/10.1016/j.neuroimage.2018.10.040

70 George, A. S., Baskar, T., & Srikaanth, P. B. (2024). The erosion of cognitive skills in the technological age: How reliance on technology impacts critical thinking, problem-solving, and creativity. *Partners Universal Innovative Research Publication (PUIRP), 2*(3), 147. https://doi.org/10.5281/zenodo.11671150

71 AI Tools: https://aitoolsguide.com/
Creative Bloq: https://www.creativebloq.com/
Medium: https://medium.com/
Awwwards: https://www.awwwards.com/awwwards/collections/ai-tools-for-designers/
GPT Store: https://chatgpt.com/gpts

72 Howarth, J. (2024, May 10). Worldwide daily social media usage (New 2024 data). *Exploding Topics.* https://explodingtopics.com/blog/social-media-usage

73 Statista Research Department. (2024, May 22). Average hours per day spent on socializing and communicating by the U.S. population from 2009 to 2022. *Statista.* https://www.statista.com/statistics/189527/daily-time-spent-on-socializing-and-communicating-in-the-us-since-2009/

74 Guttmann, A. (2024, June 20). Average daily media use in the United States in 3rd quarter of 2023, by medium. *Statista.* https://www.statista.com/statistics/1289938/daily-media-usage-us/

75 Tahmaseb-McConatha, J. (2022, October 19). Technology use, loneliness, and isolation. *Psychology Today.* https://www.psychologytoday.com/us/blog/live-long-and-prosper/202210/technology-use-loneliness-and-isolation

76 Chow, A. R. (2023, February 23). AI-human romances are flourishing—And this is just the beginning. *Time.* https://time.com/6257790/ai-chatbots-love/

77 Ballard, J. (2024, April 9). Americans' top feeling about AI: Caution. *YouGov*. https://today.yougov.com/technology/articles/49099-americans-2024-poll-ai-top-feeling-caution

78 Metze, K., Morandin-Reis, R. C., Lorand-Metze, I., & Florindo, J. B. (2023). The amount of errors in ChatGPT's responses is indirectly correlated with the number of publications related to the topic under investigation. *Annals of Biomedical Engineering, 51*(7), 1360-1361. https://doi.org/10.1007/s10439-023-03205-1

79 Magesh, V., Surani, F., Dahl, M., Suzgun, M., Manning, C. D., & Ho, D. E. (n.d.). Hallucination-Free? Assessing the reliability of leading AI legal research tools [PDF file]. Stanford University. hhttps://dho.stanford.edu/wp-content/uploads/Legal_RAG_Hallucinations.pdf

80 Schwab, K. (n.d.). The Fourth Industrial Revolution. World Economic Forum. https://www.weforum.org/about/the-fourth-industrial-revolution-by-klaus-schwab/

81 McKinsey & Company. (2024, July 3). Gen AI casts a wider net. McKinsey & Company. https://www.mckinsey.com/featured-insights/sustainable-inclusive-growth/chart-of-the-day/gen-ai-casts-a-wider-net

82 Magnifi. (2023, August 23). Which industries are adopting AI the most and least? *Magnifi.* https://magnifi.com/learn/which-industries-are-adopting-ai-the-most-and-least

83 World Economic Forum. (2024, January 9). More and more jobs can be done from anywhere. What does that mean for workers? World Economic Forum. https://www.weforum.org/agenda/2024/01/remote-global-digital-jobs-whitepaper/

84 Gregg, I. (2024, March 12). The future of work: Embracing AI's job creation potential. Forbes. https://www.forbes.com/sites/forbestechcouncil/2024/03/12/the-future-of-work-embracing-ais-job-creation-potential/

85 Coursera: https://www.coursera.org/
Google for Educators: https://grow.google/ai-for-educators/
Class Central: https://www.classcentral.com/

86 Anderson, J. (2023, February 9). Educating in a world of artificial intelligence. Harvard Graduate School of Education. https://www.gse.harvard.edu/ideas/edcast/23/02/educating-world-artificial-intelligence

87 Dergaa, I., Ben Saad, H., Glenn, J. M., Amamou, B., Ben Aissa, M., Guelmami, N., Fekih-Romdhane, F., & Chamari, K. (2024). From tools

to threats: A reflection on the impact of artificial-intelligence chatbots on cognitive health. *Frontiers in Psychology*, 15, 1259845. https://doi.org/10.3389/fpsyg.2024.1259845

88 Lumosity: https://www.lumosity.com/en/

89 Sparks, A. (2024, July 30). AI detection software doesn't work. Do this instead. *Ditch That Textbook*. https://ditchthattextbook.com/ai-detection-software-doesnt-work-do-this-instead/

90 *EdApp* (SC Training): https://training.safetyculture.com/blog/our-next-chapter/

91 *Canva for Educators*: https://www.canva.com/education/

92 *Synthesia*: https://www.synthesia.io/

Pictory: https://pictory.ai/?el=2000&htrafficsource=pictoryblog

93 *Breakout EDU:* https://breakoutedu.com/

94 *Hello History*: https://www.hellohistory.ai/

Duolingo Max: https://blog.duolingo.com/duolingo-max/

Querium: https://www.querium.com/

Plaito: https://kodora.ai/ai-tool/plaito/

SoloLearn: https://www.sololearn.com/en/

Khanmigo: https://www.khanmigo.ai/

95 *Fetchy:* https://www.fetchy.com/

96-97 Graylin, A. W., & Rosenberg, L. (2024). *Our next reality: How the AI-powered metaverse will reshape the world*. Quercus. Kindle Edition.

98 AI.gov: https://ai.gov/

MIT Technology Review: https://www.technologyreview.com/topic/artificial-intelligence/

The Organisation for Economic Co-operation and Development (OECD): https://www.oecd.org/

Made in the USA
Monee, IL
03 November 2026

67379474R00049